The Life Cycle of a

GRASSHOPPER

Jill Bailey

Illustrated by
Carolyn Scrace

Reading Consultant:
Diana Bentley

Life Cycles

Cover illustration: Jackie Harland
Editor: Janet De Saulles

First published in 1989 by
Wayland (Publishers) Limited
61 Western Road, Hove
East Sussex, BN3 1JD, England

British Library Cataloguing in Publication Data
Bailey, Jill
   The life cycle of a grasshopper.
   1. Grasshoppers
   I. Title   11. Scrace, Carolyn, 1956–
   III. Series
   595.7'26

   ISBN 1–85210–772–3

Typeset in the UK by DP Press Limited, Sevenoaks, Kent
Printed and bound by Casterman S.A., Belgium

Notes for parents and teachers
Each title in this series has been specially written and
designed as a first natural history book for young readers.
For less able readers there are introductory captions,
while the more detailed text explains each illustration.

Contents

All the words that are
in **bold** are explained in
the glossary on page 31.

A grasshopper is an **insect**.

Look carefully at this picture. Can you see how the grasshopper's body is made up of three parts – a head, a **thorax** and a long **abdomen**? On the grasshopper's head are two eyes, a pair of feelers for touching and smelling, and a mouth. The thorax holds three pairs of legs and two pairs of wings. The abdomen contains the grasshopper's stomach.

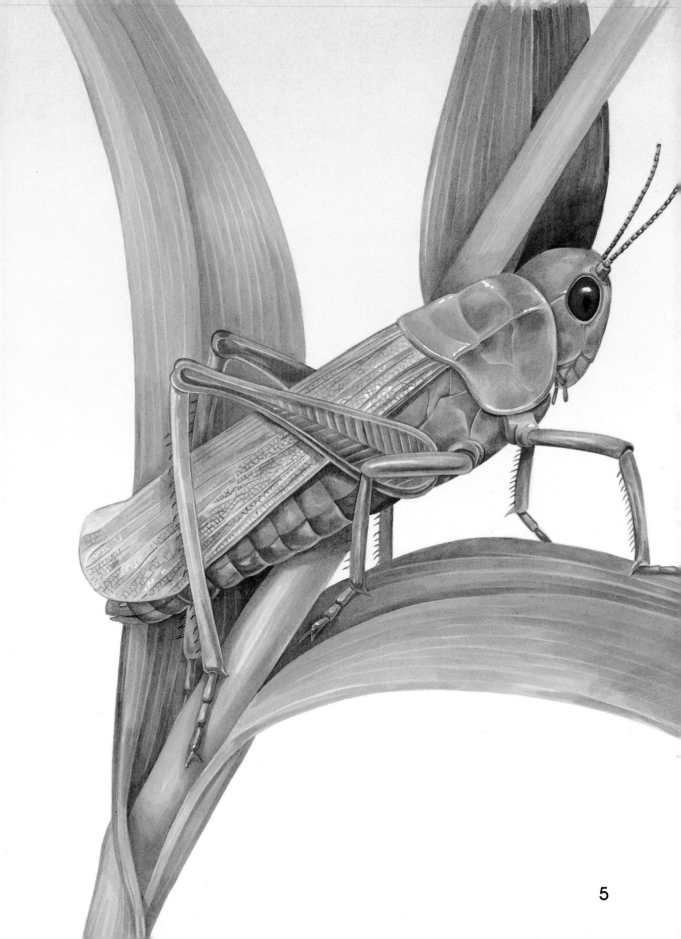

5

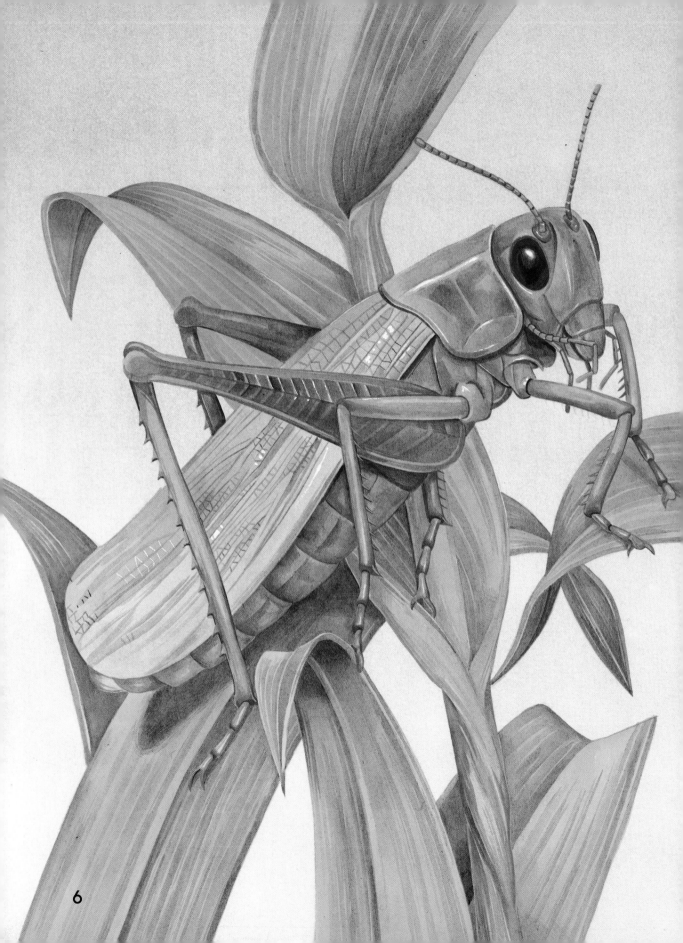

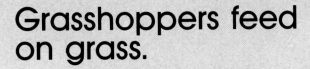

Grasshoppers feed on grass.

The grasshopper's mouth has two large, horny lips. In between is a pair of sharp jaws called **mandibles**. These have saw-like edges for cutting up food. Behind them is another pair of jaws which carry tiny feelers for tasting the food.

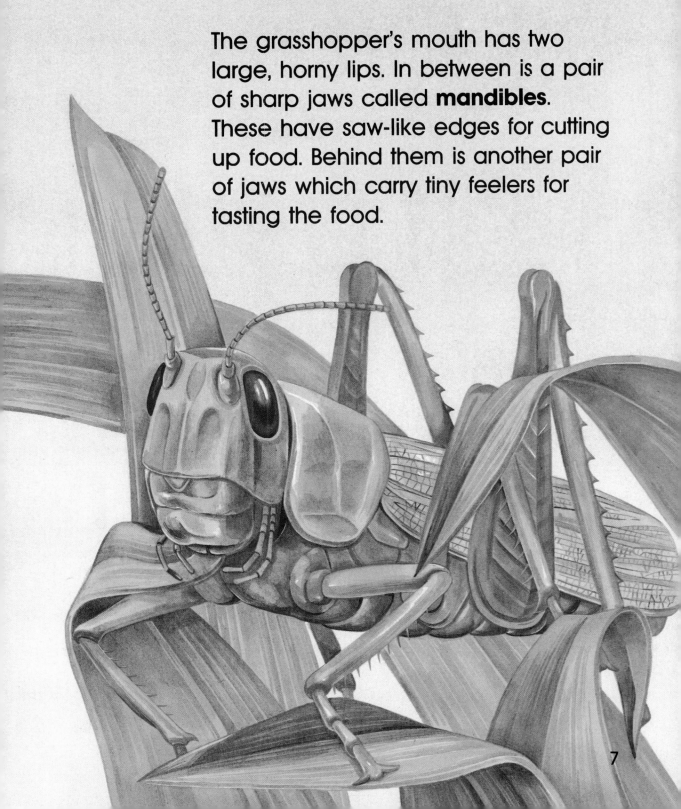

Grasshoppers are good at jumping and flying.

The grasshopper's back legs are very long and powerful for jumping. As they jump, grasshoppers spread their wings to help them glide through the air. The back wings are very large and thin. When they are not being used, they are safely folded up under the tough front wings.

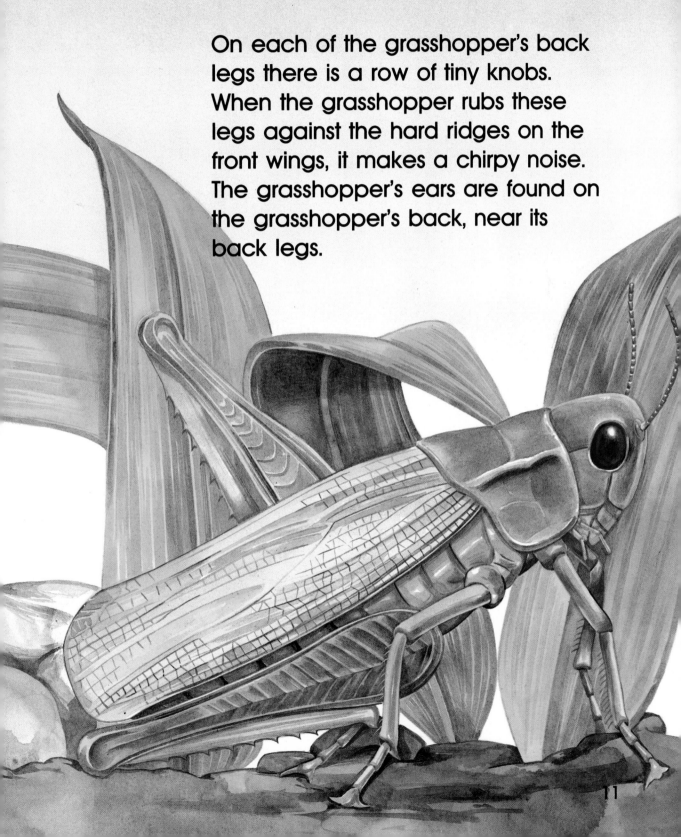

Grasshoppers sing.

On each of the grasshopper's back legs there is a row of tiny knobs. When the grasshopper rubs these legs against the hard ridges on the front wings, it makes a chirpy noise. The grasshopper's ears are found on the grasshopper's back, near its back legs.

The male grasshopper sings to find a female.

Grasshoppers cannot see each other very well in the long grass, so they call to each other instead. The male grasshopper sings to warn other males to keep away from his patch of grass. He sings a different song to attract a **mate**. She answers with a much fainter song as she crawls towards him.

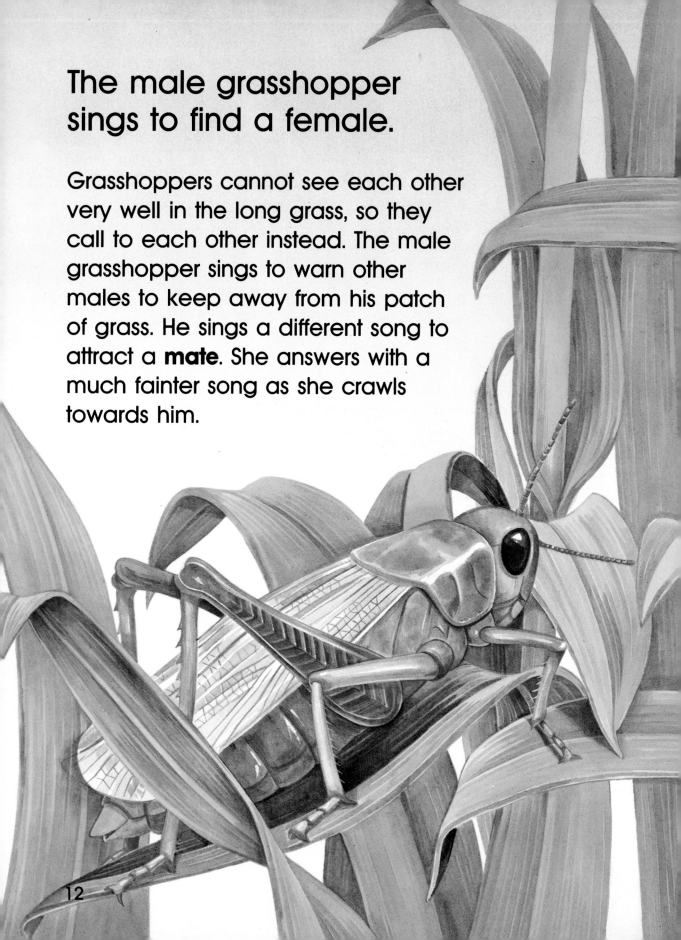

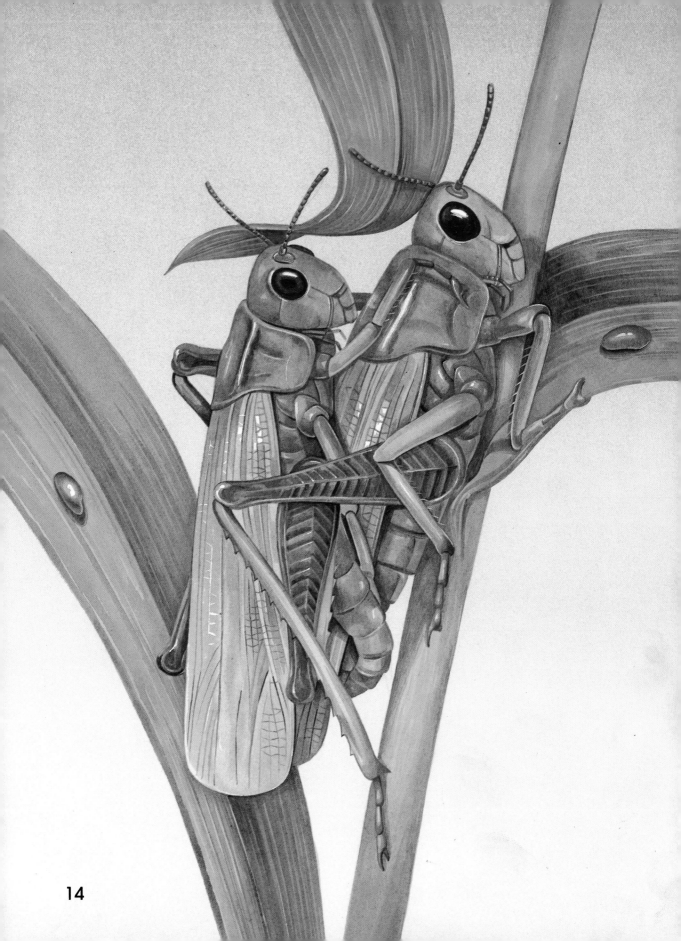

The male and female grasshoppers **mate**.

The female's eggs are stored inside her abdomen. She clings tightly to a grass stalk while the male climbs on her back. Then he squirts a special liquid into her abdomen to make the eggs ripen. A pair of grasshoppers can stay locked together for several hours while they mate.

The female grasshopper lays her eggs.

After a few days the female's eggs are ripe. She pushes the tip of her abdomen into the ground and squeezes out the eggs. She swallows air to blow up her abdomen so it can push deeper into the soil. Then she covers the eggs in white froth. This soon goes hard, and makes a case to protect the eggs.

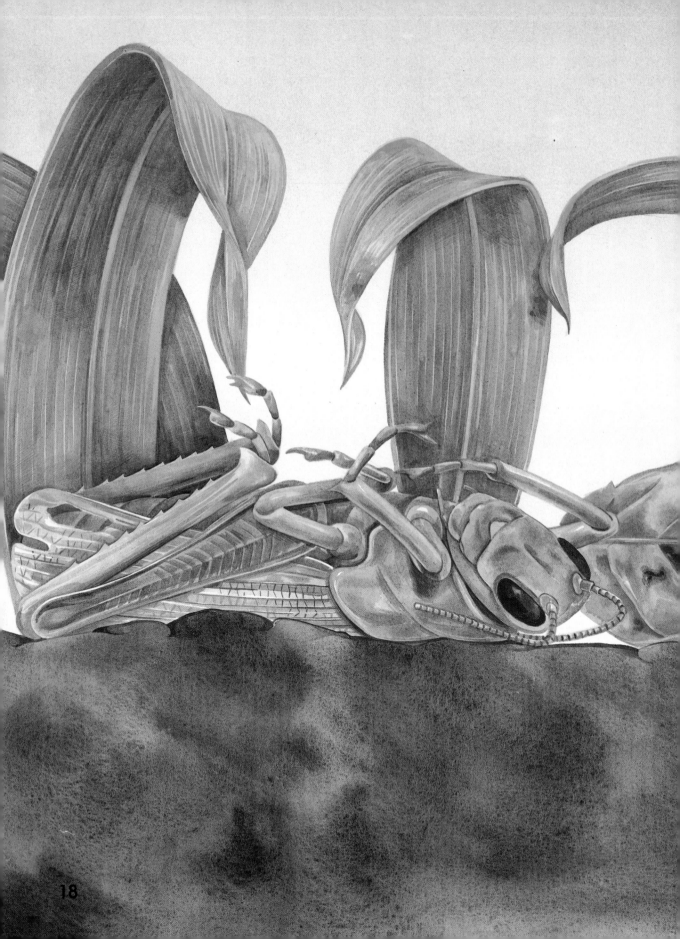

The adult grasshoppers die in the autumn.

Grasshoppers cannot survive the cold, so they live for only one summer. But the eggs live through the winter in the soil, protected by the egg case. The frost and snow cannot reach them deep in the ground. They will **hatch** into new grasshoppers when the warm weather comes.

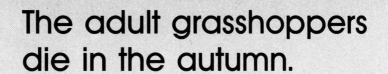

In spring the eggs hatch.

When the eggs hatch each new grasshopper looks like a tiny, fat worm. It has no legs, feelers or wings. But after a few minutes, its skin splits, and a baby grasshopper crawls out. The baby grasshopper is called a **nymph**. It looks just like its parents, but it doesn't have any wings.

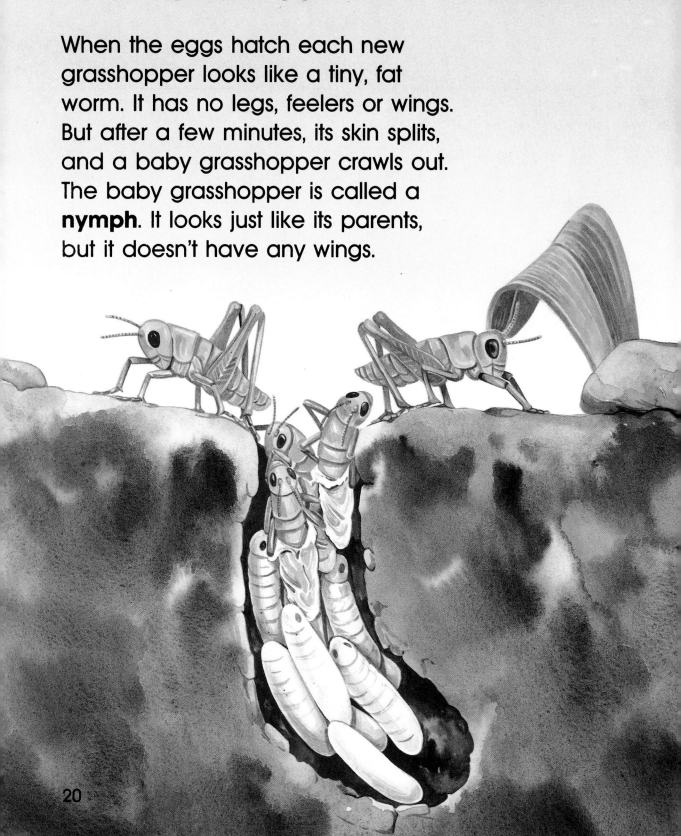

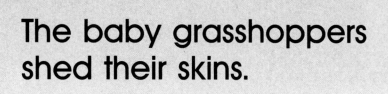

The baby grasshoppers shed their skins.

The skins of the young nymphs do not grow. Instead the young grasshoppers have to shed their skins as they grow bigger. They do this five times before they are fully grown. This is called **moulting**. After each moult the wings grow a little bigger.

The young grasshoppers have grown their wings.

At last the young grasshoppers are fully grown. Their wings are big and strong and they can fly as well as jump. Now they are ready to look for mates.

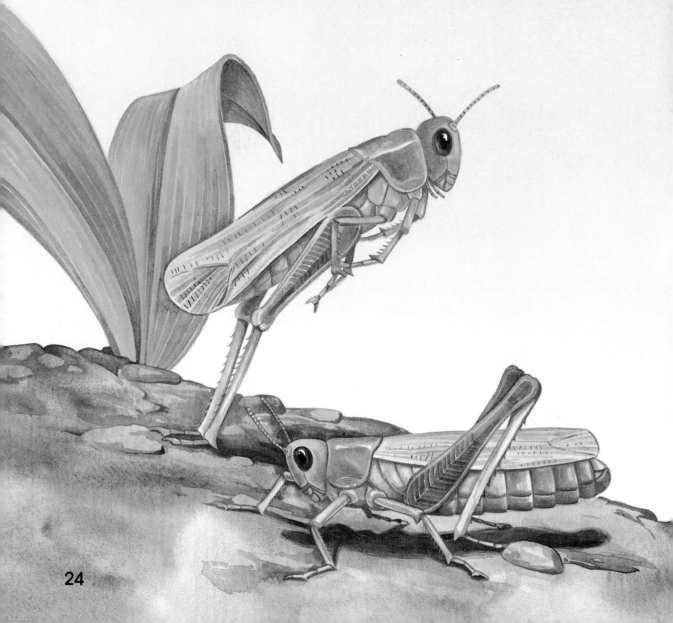

Grasshoppers have many enemies.

Frogs and toads like to eat grasshoppers, catching them on their sticky tongues. Birds and cats also hunt them. They must even watch out that they do not leap into spiders' webs. The grasshoppers' green and brown colours make them difficult to see among the grasses, but as soon as they move, their enemies can spot them. Then they must leap away as quickly as they can.

Looking at grasshoppers.

Grasshoppers live in grassy places. You can find them by listening for their songs. There are many different kinds, each with its own special song. You can keep grasshoppers in a large glass tank, such as an old aquarium. Put in some soil and grass, and a few stones and large sticks for shelter. They will need a little tray of water to drink, and some leaves to eat. Cover the top with netting to prevent them jumping out.

Here are some different kinds of grasshoppers.

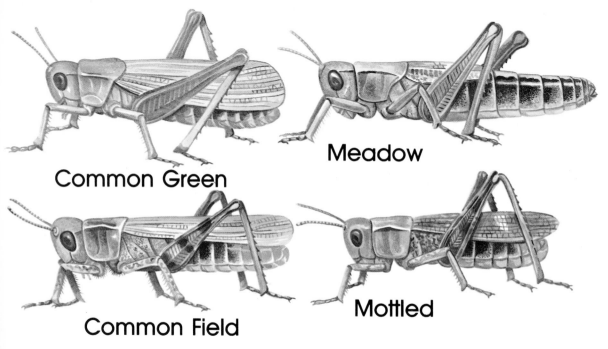

Common Green

Meadow

Common Field

Mottled

The life cycle of a grasshopper.

How many stages of the life cycle of a grasshopper can you remember?

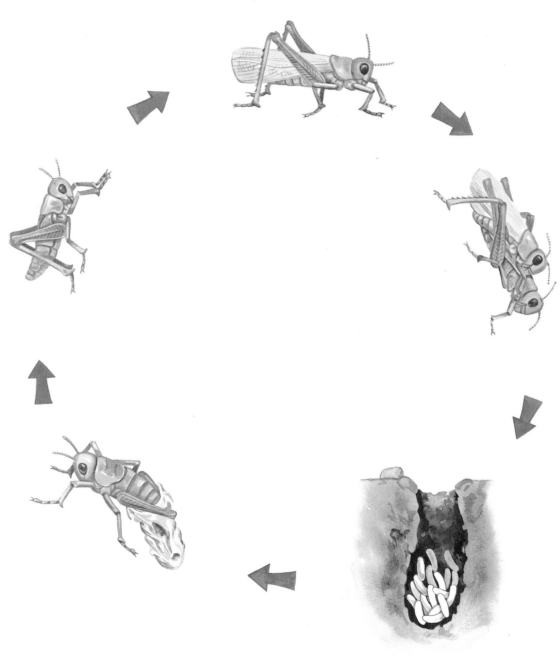

Glossary

Abdomen The rear part of an insect's body.

Hatch To break out of an egg.

Insect A small animal with six legs and two pairs of wings. Its body is divided into three parts – head, thorax and abdomen.

Mandibles The sharp biting jaws of an insect.

Mate (noun) One of a pair of grasshoppers that have come together so that the male can fertilize the female's eggs.

Mate (verb) This is when the male (father) and female (mother) animals join together. It is how a baby animal is started.

Moulting This is when the nymph gets rid of its skin. The nymph has grown too big for it. It needs a new skin so that it can go on growing.

Nymph A baby insect that looks very similar to its parents.

Thorax The middle part of an insect's body which bears the legs and wings.

Finding out more

Here are some books to read to find out more about grasshoppers and crickets.

Discovering Crickets and Grasshoppers by Keith Porter (Wayland, 1986)
Grasshoppers by Jane Dallinger (Lerner, 1983)
Insects by Anthony Wootton (Franklin Watts, 1980)
Spotter's Guide to Insects by Anthony Wootton (Usborne, 1979)
The Grasshopper by Gunilla Ingves (A. & C. Black, 1983)

Index